How Worm

Became Everything

By Christian Darkin

ISBN 978-1-9998930-1-9

First Printing 2019
Published by
Rational Stories
www.RationalStories.com

There was once a worm.

The worm wanted to swim...

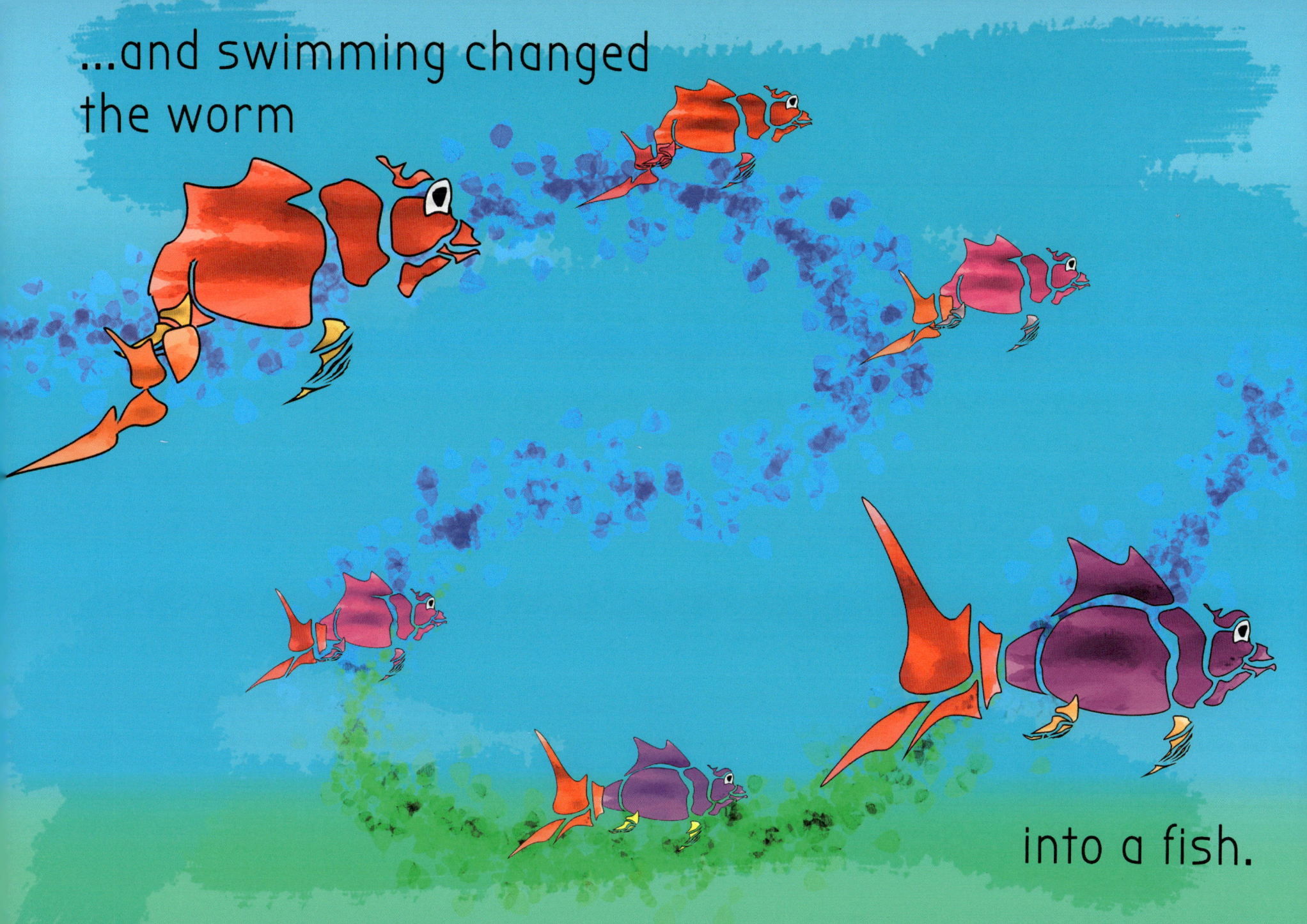

...and swimming changed the worm

into a fish.

The fish wanted
to crawl...

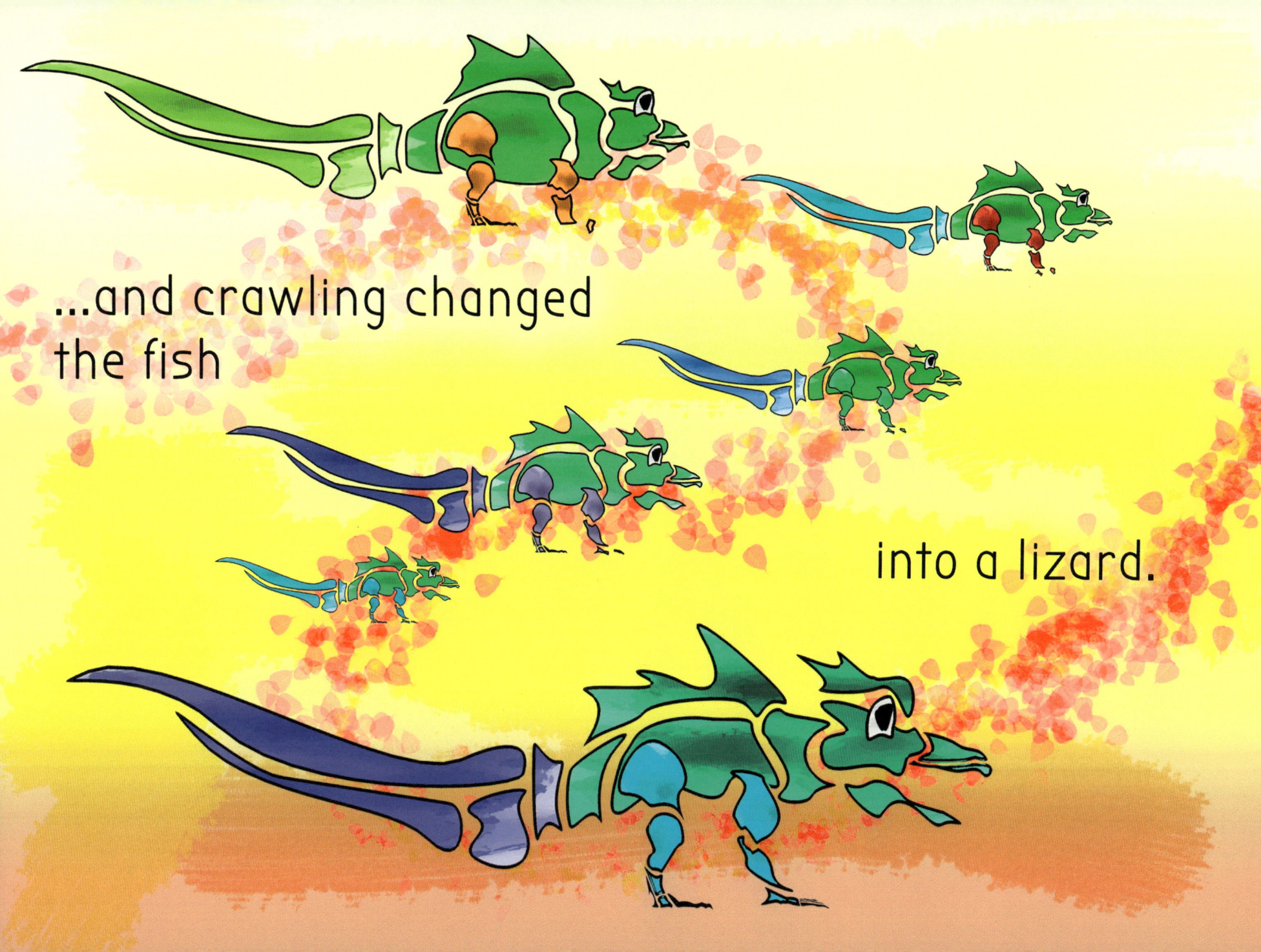

...and crawling changed the fish

into a lizard.

The lizard wanted to run...

...and running changed the lizard

into a dinosaur.

The dinosaur wanted to fly...

... and flying changed the dinosaur

into a bird.

There was
once a
worm.

The worm wanted
to crawl...

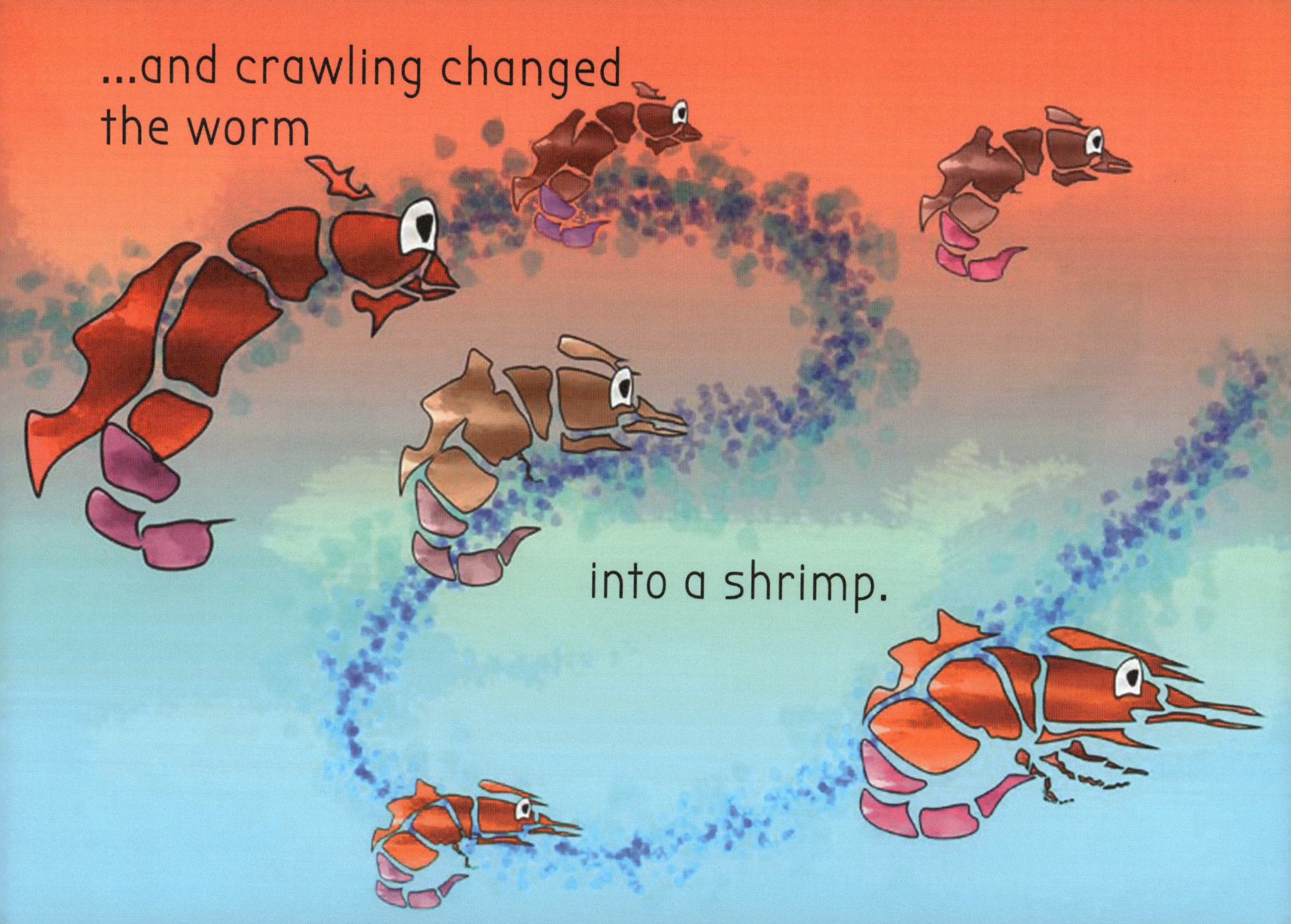

...and crawling changed
the worm

into a shrimp.

The shrimp wanted to march...

The insect wanted
to fly...

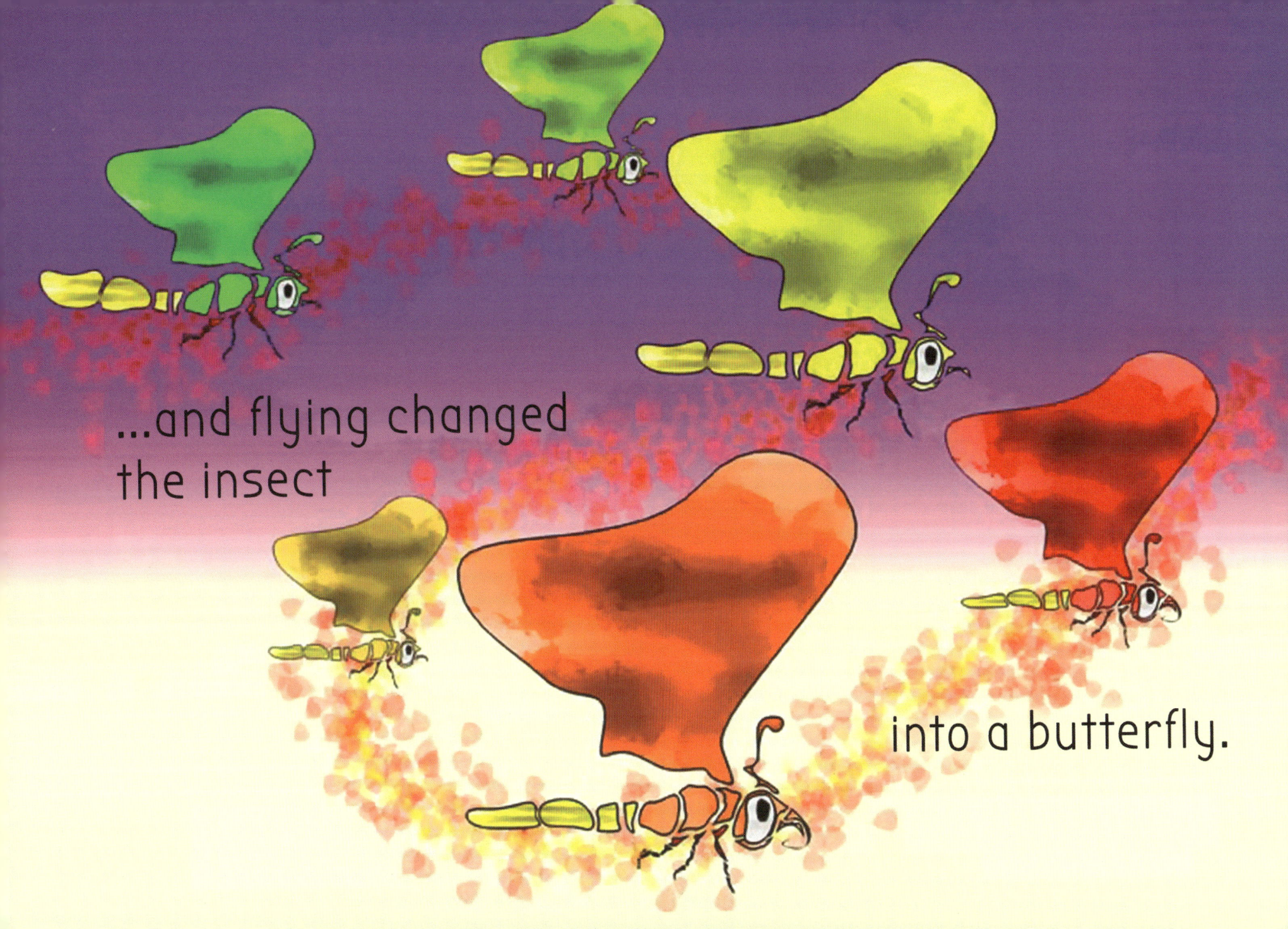

...and flying changed the insect

into a butterfly.

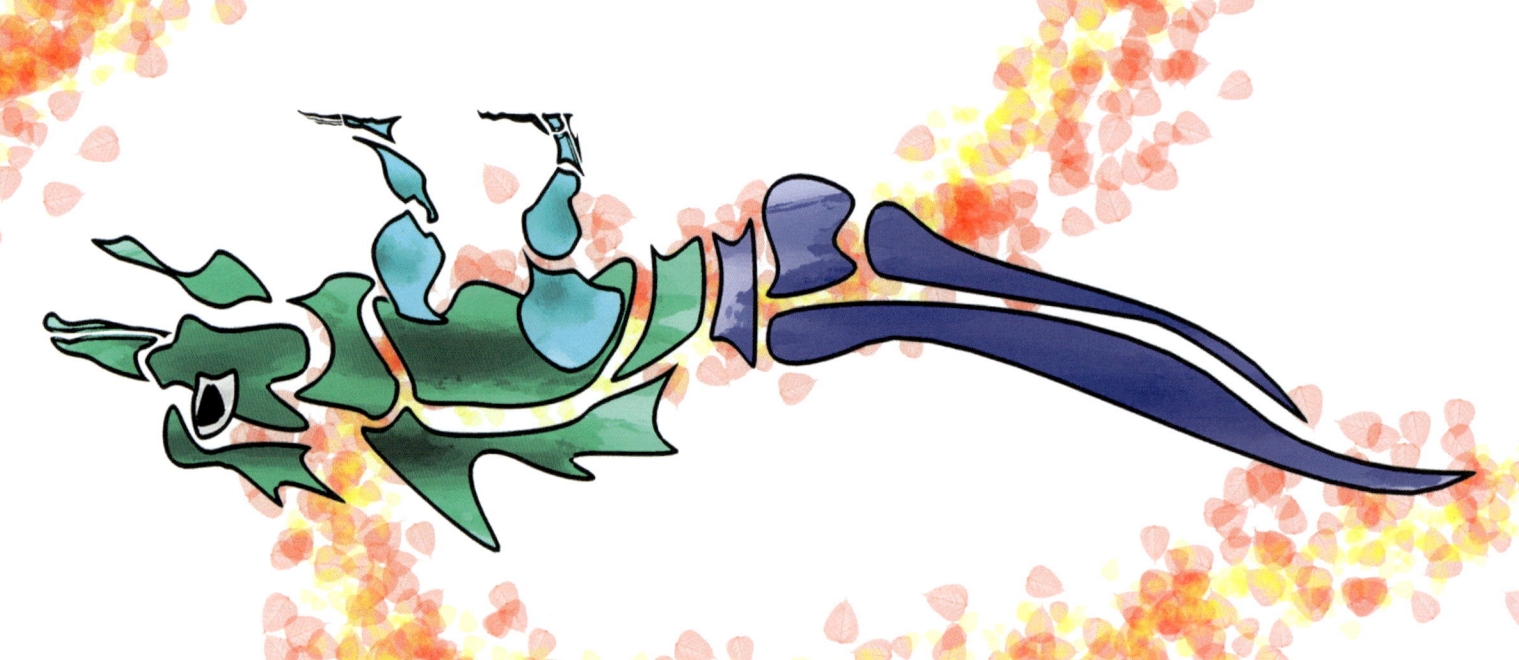

There was
once a
lizard.

The lizard wanted to keep warm...

...and keeping warm changed the lizard

into a mouse.

The mouse wanted to jump...

The lemur wanted
to climb...

The ape wanted
to walk...

What do you want to do now?

Printed in Great Britain
by Amazon